WITHDRAWN

YOU CALL, WE HAUL

Tana Reiff

American Guidance Service, Inc.
Circle Pines, Minnesota 55014-1796
1-800-328-2560

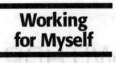

Working for Myself

Beauty and the Business
Clean as a Whistle
Cooking for a Crowd
Crafting a Business
The Flower Man
The Green Team
Handy All Around
Other People's Pets
Your Kids and Mine

Cover Illustration: James Balkovek
Cover Design: Ina McInnis
Text Designer: Diann Abbott

Printed in the United States of America
ISBN 0-7854-1114-3 (Previously ISBN 1-56103-902-0)
Product Number 40830
A 0 9 8 7 6 5 4 3 2

CONTENTS

Chapter

CHAPTER 1

In Need of Work

Barry wasn't the type to let things upset him very much—even big things, like when his marriage ended. He wasn't happy when he and Suzanne split up. But he had seen it coming. It wasn't a surprise. The hardest part was not living with his son. But he saw the boy often. His life was going all right.

But this was different. Now his hand was shaking as he held the pink slip of paper in his hand. Everyone at the plant had gotten these layoff notices today. Barry felt as though he had been shot in the back. How can a guy have a good job for ten years and then lose it just like that? Why didn't he see this one coming? As he left the plant, Barry asked himself these questions over and over again. It was all that he could think about the whole way home.

Barry drove his pickup into the garage. He went inside and sat down with a cold drink. He turned on the TV and just sat staring at it. The ink on the pink slip smeared as he held it through six TV shows. The paper was wet and wrinkled by the time the late night news came on. The next morning Barry woke up with a stiff neck. He had slept all night in that chair. The pink slip had fallen onto the floor beside him.

A new day. But he had nothing to do, and nowhere to go. "This is not me. I can't believe this," Barry said out loud. He hung around the house all day. He felt terrible.

The next day started out the same way. In the afternoon Barry drove his truck downtown to see about filing for unemployment compensation. He hadn't even bothered to shave by the time he picked up his son at 4:00 P.M.

"Can I take my bike along?" the boy, Chad, asked.

"Sure," Barry said. "Throw it in the back of the truck."

Barry looked at his pickup. It was his prized possession—the only thing he owned that was really worth anything. That's when the idea suddenly came to him. "What can you do when all you have left is your pickup truck?" Barry said to himself. "You can try to *make some money* with your pickup truck."

The truck was only a year old. It ran like a charm. Barry had used it to move his stuff out of the house after he and Suzanne broke up. He had moved everything he had in only three trips. So why not use the truck to haul things for other people? Why not help people move? Or carry junk to the dump? There had to be money in such a business. Besides, his favorite place in the world was sitting in the truck's cab. He'd love to have a job that meant using his truck all day.

Chad rode up front with him. Barry looked back at the bed of the truck. One bike took up almost no room at all. Just think how much stuff would fit back there if he added side rails. Barry knew he could build some in no time.

His mind was racing. He wasn't thinking about his driving.

"Aren't you going a little fast, Dad?" said Chad.

Barry's foot popped off the gas. "Sorry, buddy," he said, laughing. "I'm sure glad I have you around to keep me in line!"

Barry started thinking out loud while he and Chad were having dinner.

"So I'm going to look into this idea,"

he said. "I can haul at least three loads a day, don't you think? That would be pretty good money, don't you think? Between hauling and unemployment money, I'll do fine."

Chad really didn't understand the money talk. But he did think it would be cool for his dad to drive a truck around all day.

"I could even work as your helper on Saturdays," Chad said.

"Hey, sure thing," Barry said. "And summers, too." He looked at his young son. He didn't think the skinny little boy could be very much help. But Chad was fun to be with. He could at least ride along and keep his dad company.

Two days later Chad's mother called. "I was sorry to hear you lost your job," Suzanne said. "So you're going into business for yourself, huh?"

"I'm going to look into it, anyway," said Barry.

"You had better make sure that I still

get my child support," said Suzanne.

Barry hated hearing his ex-wife say that. "Don't worry about your child support," he said.

"That isn't really why I called," Suzanne went on in a friendlier voice. "I called because my friend Melissa is cleaning out her basement. She needs someone to haul away the junk. I told her I'd ask you."

"What's her number?" Barry asked. He wrote it down. "I'll call her. And thanks, Suzanne."

Laid off less than a week and he already had some work! Barry thought this might be the start of a good thing.

CHAPTER 2

The First Job

Barry called Suzanne's friend Melissa. He said he'd be glad to clean out her basement. When she asked how much he'd charge, he pulled a price out of the air. He really didn't know if he was asking too much or too little. But Melissa agreed right away.

"When do you want me to come over?" Barry asked.

"I'd say the sooner, the better," said

Melissa in a friendly voice.

"Is tomorrow soon enough?"

"That's fine. How about 8:00 A.M.?" Melissa asked.

"See you then," Barry said.

That afternoon Barry went down to the town hall and bought a dump permit. He knew that he would need one. No one could just go to the town dump on the spur of the moment. Everyone had to have a permit. There was a dumping fee for each load, too.

Later he stopped and bought a pair of extra-thick gloves. Then he went on to the Rent Center and rented a heavy-duty hand truck.

On the way home Barry bought some two-by-fours. When he got home, he built side rails onto the truck bed. He put his toolbox in the cab and a shovel in the back. He made sure the tire jack was there, too.

The next morning Barry drove to the gas station to fill up the truck. He also checked the truck's water, oil, and tires. Melissa was at the door waiting when Barry got to her house.

"You must be Barry," she said.

"That's me," Barry said. "I don't remember meeting you. How come Suzanne never mentioned you when she and I were married?"

"Oh, I just started working with Suzanne about six months ago," Melissa said. Barry followed her down to the basement.

"Is there an outside door down here?" he asked her. There was, but he had to move a pile of stuff to get to it. He pushed the door open. "That makes things much easier," he said. "Now I won't have to haul all this junk through your house."

Barry got to work. He carried out ten bags of trash, a pile of old wood, and some broken chairs. He used the hand truck to haul out a rusty old stove. He packed the truck carefully. He stood the big things along the sides. He filled in the middle of the truck bed with the little stuff. It took just about half the day to finish the job.

While he was working, Melissa came down to the basement a few times. Whenever Barry stopped to take a break, they talked for a few minutes. Each time Melissa had a cool drink ready for him. The two of them really seemed to hit it off.

When Barry packed the last load, Melissa paid him. "Thanks a lot," she said. "Getting that junk out of here takes a load off my mind."

Barry found himself very attracted to this woman. "You've given *my* mind a lift, too," he said with a little smile.

That took Melissa by surprise. "What do you mean?" she asked.

"I mean, would you go out with me? I'm just a hard-working guy. But I'd sure like to take you out for a nice dinner."

Melissa didn't say anything. Barry thought he knew what she was thinking.

"Oh, don't worry about Suzanne," he said. "We've been divorced for almost a year now. It's really none of her business."

"All right," said Melissa. "I'll go out with you. But do you go out with all your customers?"

"Well, I really don't know," Barry laughed. "You're my first customer!"

Melissa laughed, too. "What a way to meet men!" she said.

Barry was riding high the whole way home. "Hey! Good work and a nice woman, all in one stop!" he said out loud. "How's that for a great day!"

That night Barry worked with some numbers. He figured out how long Melissa's job had taken and how much he had charged her. For the time it took, he saw that he hadn't charged enough for the job.

Barry realized he would have to buy a few more tools. A hand truck of his own would be good to have, too. He took a guess at how much he would have to spend. He knew he needed advertising to find some customers. He called the local newspaper to find out the cost of an ad.

Finally he wrote down how much money he needed to make in a year. He divided that by 52 weeks. Then he divided the new number by five days. He'd have to haul about three loads a day to make this business work.

Barry looked at everything he had written down. Then he came up with a basic price per load. He decided that he would charge more for each extra flight of stairs. He'd also charge more if he had to drive more than five miles out of town.

Barry studied the figures he had written down. It looked like this could turn into a serious business.

CHAPTER 3

Getting More Work

Barry knew that word of mouth is a great way to get work. But he also knew that he couldn't count on *all* his work coming from one person telling another. Suzanne had gotten him the first job with Melissa. But now he had to go out and get more work on his own. Where should he begin?

He had a hundred ideas. He started with a newspaper ad. He worked on it most of the morning. He wanted every word to sound just right. And the whole ad had to fit on three lines.

He thought about what people would want. Most of all, they would want him to get rid of their trash or move them into a new place. They wouldn't want a mess in the house. They wouldn't want to pay too much. And they'd probably want the job done as soon as possible.

Barry wrote:

General hauling. Yard, basement, and attic clean-up. Fast service. Low rates. U call, I haul. 555-2883.

Late in the morning he called the newspaper and placed the ad. The next morning he opened the paper and there it was. It was in a section of the want ads called *Hauling*. Barry saw that he wasn't alone in this business. He wondered if there was enough work out there for everyone.

So he got out some paper and made six signs. He drew a picture of his truck and wrote down his name and phone number. At the top he wrote the word *HAULING* in big letters. Across the bottom he wrote his phone number sideways ten times. Then he cut the paper between each number. That way, people could just tear off a phone number to take with them.

Barry took the signs to three food stores. He asked each store's manager if he could put up his sign on the community bulletin board. They all said that he could.

Then he took the other three signs to senior citizens centers. He thought that older people would be glad to hear about his services. He knew it was too much for many of them to do such heavy work themselves.

On the way home Barry stopped at a print shop. He wrote down his name, address, and phone number. Then he ordered 500 business cards. The woman at the print shop said the cards would be ready in a few days.

Barry also stopped at a small sign shop. He wanted a sign to stick on the side of his truck. "Most people get two," the owner said. "One for each side of the truck." Barry ordered two signs.

Over the next few days he talked with a few people who might have some work for him.

One woman was the owner of a real estate company. "Give me a call when people move out of your apartments," Barry said. "I'll come and pick up any junk they leave behind."

"Sometimes we put in new carpet when someone moves out," the woman said. "Can you haul out old carpet?"

"Sure," said Barry. "Anything."

He went to see his buddy Allen, who built houses. "Who cleans up after you?" Barry asked. "You know—how do you get rid of what's left behind when you're finished building?"

"We usually call a hauler," said Allen.

"Great! I'm in the hauling business

now," Barry said to his friend.

Allen looked at Barry's truck. "Your truck is kind of small, isn't it?"

"You'd be surprised at how big a load I can carry," Barry said.

Allen said he would give Barry a call one of these days.

Barry had nothing to do that night. Then he had an idea. Every Tuesday night there was an auction downtown. He hopped in the truck and drove there. He wasn't looking to buy anything. He was looking for work.

Before the auction began, Barry talked to the auctioneer. "You would really help me out if you would tell the people I'm here," Barry said. "Tell them I'll be waiting outside. If they need someone to haul their stuff home, I'll do it."

"You might be helping me out, too," said the auctioneer. "People often worry about buying the big stuff if they don't know how they can get it home."

Sure enough, that night Barry moved

a bed and a chest of drawers.

When he got home, the phone rang. He hoped it was a call for more work. But it was Melissa.

"So how's it going?" Melissa wanted to know.

"I have ads out all over town," said Barry. "But no one's calling. I don't know what's wrong."

"How do you know that no one's calling?" Melissa asked. "I've been trying to call you all day. The phone just rings and rings."

"I haven't been home," Barry said.

"Don't you have an answering machine?"

"I think there's an old one around here somewhere. But I'm not sure it even works anymore," Barry said. "Boy, am I stupid! I put an ad in the paper. I have signs up in stores and senior centers. I have real estate people all lined up. And no one is home if people call."

"Well, I just called to tell you I had a

great time Saturday," said Melissa.

"Me, too," said Barry. "Did you tell Suzanne?"

"Let's keep this our little secret for now, OK?" said Melissa.

"Sure," said Barry. "There's no reason why Suzanne needs to know."

"One more thing," Melissa added. "Get an answering machine!"

CHAPTER 4

A Setback

The next morning Barry went out and bought a new answering machine. When he got back home the phone began to ring off the hook. Every time Barry hung up, it rang again. He learned that most callers wanted him as soon as possible. He lined up three jobs before 10:00.

Barry wrote down the three addresses. He told each person what time he thought he would get there. Then, before he left, he hooked up the new answering machine.

The sky was gray today. The paper said it might rain. Barry threw one large and one small plastic tarp in the truck bed. He wanted to be ready, just in case.

He headed for the first job out on Owl Hill Road. Barry had heard the name of the street, but he had never been there. All he knew was that it was outside of town.

Barry drove and drove. He stopped three times to ask directions. He wasted 45 minutes and a lot of gas. "Next time, I'll ask directions on the phone," he laughed to himself. "And it wouldn't hurt to carry a map, either."

Barry finally found the house. But getting lost was going to make him run late all day.

At the first job he cleaned out a yard. It took two trips to the dump. The sky was dark but the rain held off.

At the second place he hauled out an

old washing machine. He was glad he had bought his own heavy-duty hand truck. It made the job easy. And he was glad that he didn't have to put out money to rent one.

Barry stopped to eat lunch. He looked at the sky. There were plenty of dark clouds, but still no rain.

At the third stop Barry had to move someone out of a small apartment. He didn't get there until 2:00. By that time the rain had started. He had a whole room to load up and it was raining! It was windy, too.

"Well, I guess no one ever said it would be easy," Barry said to himself.

He laid the large tarp down on the truck bed. He put the extra part up over the cab. Then he laid the small tarp on top of the large one.

A middle-aged woman met him at the door. Barry stepped inside the one-room apartment. The place was filled with boxes. There was also a bed, a large TV, and some tables and lamps.

"I see that it's raining hard," the woman said. "I hope you can keep my things dry."

"Don't worry," Barry said. "I came prepared."

He put the woman's things under the small tarp to keep them dry. He put heavy things toward the front of the truck bed. He packed the mattress and box springs standing on their sides.

When the truck was full, Barry pulled the large tarp off the cab. He stretched it over the load and tucked the end under some boxes in the back. "No wind is going to blow this baby off," he said.

He unloaded everything at the new apartment. Then he went back and forth again for two more loads.

He worked until dark. By that time he was soaking wet. His face stung from the wind. And his back was starting to hurt. He should have had some help lifting that mattress and box spring. Not that the bed was too heavy—but it was too big for one guy to handle by himself.

The woman smiled as she wrote out a check to Barry. "You earned every cent of it. Thank you so much for all your hard work," she said.

Barry went straight home. The answering machine was blinking away. He was sure people had called with more work. But at that point he was too tired to listen to his messages. And his back was beginning to bother him—a lot. He made himself a sandwich and took a long, hot bath.

Then, before he could call anyone back, he fell sound asleep.

Barry woke up early the next morning. His first thought was to return the calls from the day before. But, when he tried to get up, he cried out in terrible pain. It felt like someone had stuck a knife in his back.

"Oh, my God," he said to himself. "How will I be able to do any work today? I can't even get out of bed."

Barry knew right away that he was injured. Slowly, and with a lot of pain, he crept out of bed. He went over to the phone and called Melissa at her job. He told her what had happened. She told him to call his doctor. Then she offered to take some time off to drive him to the doctor's office.

When Melissa got there, Barry was already dressed. But he could barely walk without crying out in pain.

"I don't know what I did," Barry said. "I didn't feel this bad yesterday—even after I finished that last job."

"Let's see what the doctor says," Melissa said as she helped him out the door. Then they drove off in her car.

"It's a bad sprain," the doctor said. "You pulled a muscle, too. But you're lucky. From what you've told me, it could have been a lot worse."

"I don't *feel* lucky," Barry said. "How long before it gets better?"

"A week, maybe a little longer," the doctor said. "Rest is what you need. And I'll give you some medicine. But the most important thing is to stay off your feet as much as possible. And of course, no lifting—of *anything*."

C H A P T E R   5

Second Thoughts

On the way home, Barry was upset. "What am I going to do?" he asked Melissa. "I can't work for at least a week, maybe longer. I'll lose jobs from people who need hauling now. And I won't have any money coming in."

"It can't be helped," Melissa said. "You heard the doctor. Besides, if you injure yourself again, you could wind up in the hospital. If you're not careful, you could be laid up for months."

"Please, don't even mention that," Barry said. "When I lost my job at the plant, I lost my medical insurance. It's bad enough that I'll have to pay for today's doctor visit. A stay in the hospital would break me."

"Well, maybe you should get some medical coverage," Melissa said. "My neighbor is self-employed. He has basic coverage for hospital, surgery, things like that."

"Isn't it expensive?" Barry asked.

"Well, it's sure not cheap," Melissa admitted. "But it's better than paying full hospital costs."

"You know, maybe being in business for myself isn't such a good idea after all," Barry said. "At the plant at least I got paid when I was out sick. And the company paid for most of my health insurance."

After they got back to Barry's house,

Melissa helped make him comfortable on the couch. But his mood hadn't gotten any better.

"I don't know, Melissa. Maybe as soon as I get better, I should start looking for full-time work."

"Look," Melissa said. "It's natural to have second thoughts about a big step like this. Why don't you take it easy for a week? Just rest and let your back get better. And do some thinking. What will really make you happy?

"Are these problems really too big to solve? Maybe you just need to get help for some of the heavy jobs. You can hire people who won't cost too much. And I'll be sure to ask my neighbor about his health insurance."

Barry took Melissa's advice. For a week he took it easy and rested his back. Each day he got calls for work. He turned down the jobs that needed to be done right away. But he was able to schedule some jobs for the next week. He only hoped that he'd feel better by then.

Melissa came over every night to see him. She helped him work out a new price list. She gave him information about health insurance for self-employed people. They talked a lot and laughed a lot. Some nights she stayed for dinner. Barry couldn't remember the last time he had been so happy in a woman's company.

By the end of the week, Barry's mood had improved. And so had his back. He decided he would try to take a few jobs the following week. And he made one more decision.

"I'm going to give this hauling business a real try," he told Melissa that Saturday night. "I can't quit now, just because of a few problems."

"That's great!" Melissa said to him with a great big smile.

"I'm going to hire some help for the heavier jobs. And I already signed up for the same kind of health insurance that your neighbor has."

"I think you're on the right track," Melissa said.

"Thanks for all of your help and encouragement," Barry said. He held her hand tightly.

The following Monday Barry called a woman he had talked to the week before. He told her he felt well enough now to take her job. But first he needed more details.

"As I told you last week, I have an attic full of junk," the woman said. "I want everything out."

"What are the biggest items you have?" Barry asked.

"An old sofa and a bed," said the woman. "A few broken chairs. The only other things are boxes and a worn-out old carpet. Stuff like that. Nothing too heavy."

"Is there anyone who can help me with the sofa and bed?" Barry asked.

"My neighbor can help."

"Is everything packed?"

"Everything's packed."

"How many flights of stairs lead up to the attic?" Barry asked.

"Two."

"Where do you live?"

"Eight miles north of town."

Barry worked out a price. The job sounded like one load. He added five dollars a load for being more than five miles away. He added another five dollars for the second flight of stairs. The woman said the price was fine. Then Barry looked at the map and asked for directions.

He went to the attic-cleaning job the next day. He kept the directions in front of him as he drove. He found the place without any trouble.

The neighbor helped carry the heavy furniture. Barry did the rest himself. Nothing seemed to be too heavy. But he kept knocking against the wall because the stairs were so narrow. He slowed down a bit. He knew that he would have to be careful not to hurt his back again.

Barry heard something rattle inside one of the boxes. He couldn't help but wonder what was in there. "You're sure this is all junk?" he asked the woman.

"Oh, yes, it's junk," said the woman. "All of it's stuff that was here when I moved in."

"You're *sure* you want to get rid of it?" Barry asked again.

"Yes, I'm sure," she said. "I don't want any of it."

Barry had to pack the truck tightly to fit everything in one load. He started driving toward the dump. But he couldn't stop thinking about what might be in those boxes.

All of a sudden he pulled over. He got out and walked around to the back of the truck. He opened one of the boxes with his pocket knife.

At first all he could see was yellow newspaper. Then he noticed that the paper was wrapped around small things. Gently he picked up something round. The newspaper crumpled around it was dated 1942. There was a story about the Second World War.

The paper tore easily. That's when Barry saw the beautiful china sugar bowl. It looked very old, but it was in perfect shape.

He opened more boxes. There were beautiful plates in one box. There were cups and saucers in another box. It was a complete set of fine old china.

Barry headed for home instead of the dump. He drove slowly the whole way. He kept thinking how crazy it was to carry fine china in the bed of a pickup truck.

As soon as he walked in the door he called Melissa. "Can you come over here right after work?" he asked her. "I want you to look at something."

By the time she got there Barry had unwrapped all the china. He had lined the pieces up in rows across the living room floor. There were 50 pieces in all.

Melissa dropped to her knees and picked up a dinner plate. "This is lovely," she said. "My great aunt had a set of china like this. When she died my cousin got a lot of money for it."

"I don't know what to do," Barry said. "The woman told me two times that she didn't want any of it."

"Then it's yours," said Melissa.

"What do *I* want with fine china?" said Barry. "Do you want it?"

"I'd love it," Melissa said. "But I can't just *take* it. I'll tell you what. Let me find out how much it's worth. If I can afford it, I'll buy it from you."

"Maybe I'll give you a special price," Barry said with a wink.

Melissa turned the plate upside down. She wrote down the name of the china. Then she made a list of all the pieces in the set.

Barry went out to bring in more boxes.

Before long the living room floor was covered with "treasures." There were glasses and vases and knick-knacks of all sorts. "Gee, and I thought I was a *junk* hauler!" Barry laughed.

"I'd say you made out like a bandit today," Melissa said.

On his next trip to the truck, Barry took a better look at the sofa and chairs. The sofa's legs were a little scratched up, but otherwise it was in pretty good shape. "Someone would pay 50 dollars for this thing," Barry thought to himself. With some glue and a little varnish, the chairs would be fine, too.

So that weekend Barry fixed up the furniture and took it to the auction house. He came home with 100 dollars in his pocket.

Melissa found out that the china was worth at least 500 dollars. Barry sold it to her for 250 dollars.

He had made a total of 350 dollars extra from one load.

"Not bad for a day's work," Barry said to Melissa with a big smile.

"Not bad at all," Melissa agreed. She gave his hand a squeeze.

CHAPTER 6

Help Needed

After that Barry kept his eye out for valuable things on every job. There were plenty of moving jobs. But most of what he saw on hauling jobs was true junk. He soon became friends with the guys at the dump.

On the way back from the dump one day, he stopped by to see Melissa. She was cooking dinner, and the table was set with the good china. She was surprised to see Barry at the door.

"Someone special coming for dinner?" he asked.

"Actually, I'm having Suzanne and Chad for dinner," Melissa said. "I thought it was time for me to tell them about you and me."

"OK," Barry said.

"Maybe I should have talked with you first," said Melissa. "But I wanted to do this myself. Do you want to stay for dinner?"

Barry was tired and needed a shower. "No, thanks," he said. "But you can give Chad a message for me. Tell him that I'll call him later this evening. I'd like him to come along with me on a big job next Saturday."

Suzanne was not at all surprised to hear about Barry and Melissa. "He's a good guy, really," Suzanne said. "And he's always been a hard worker. If he wasn't my ex-husband, I would have tried to match the two of you up myself!"

Chad didn't take the news so well. It wasn't that he didn't like Melissa. He just didn't know her very well. The message from his dad made him feel better, though.

Barry had another moving job on Saturday. There were three rooms of furniture. And there was a small piano. Barry had never moved a piano. He decided to take along some planks to make a ramp. Still, he knew he couldn't move even a small piano by himself. After what had happened with his back, he wouldn't even try.

Then Barry had an idea. "Who are the strongest men you can think of?" he asked his son the next day.

"Weight lifters!" Chad answered.

"And where would you find weight lifters?"

"At the gym!"

"Right!"

Barry took Chad along with him to the downtown gym. They watched some men grunting as they lifted huge weights. Their muscles looked bigger than the weights. There were quite a few women there, too.

"This is cool," said Chad.

Barry asked around and quickly found three guys to help out on Saturday. He said he would pay them a flat fee for each load. "I'm not taking anything out for taxes," he told the guys. "That's up to you." He said he needed help for about an hour—only for the heavy furniture and the piano. He and Chad would be able to do the rest.

Barry and Chad met the three weight lifters at the job on Saturday. Chad carried a lot of the lighter things, like small boxes and lamps. Barry and the weight lifters got the furniture out in no time. Barry had to laugh at himself. He had always thought of himself as strong. But he was *amazed* at his helpers' strength. They made everything they did look so easy!

"Gee, Dad," said Chad. "Maybe *you* should start lifting weights."

"Maybe I should," laughed Barry. "These big strong guys are making me look like a wimp."

Then it came time to move the piano. Barry laid the planks over the front steps to make a ramp. Together the four men easily got the piano down the ramp. Then Barry moved the planks to make a ramp into the truck. Together the men slid the piano into the truck.

"I want all three of you to ride along with the piano," said Barry. "You hold onto that baby with all you've got. You have to be ready if it starts to roll. I don't have insurance to cover a smashed piano!"

The fact was that Barry had no business insurance at all. No company would cover him until he had been hauling for at least a year. High risk, they told him. So he was being extra careful with the piano.

The weight lifters were worth every penny. The piano didn't roll an inch. It reached its new home without a scratch. Thanks to the weight lifters, the whole job was a breeze.

Back at the apartment Barry paid the weight lifters and thanked them for their help. He saw that there were still some trash bags, a lamp, and a refrigerator to get rid of. "What do you want me to do with this stuff?" Barry asked the couple.

"Can you take it all away?" they said. "It's just trash. The lamp is broken. And we're getting a new refrigerator in the new place."

Chad carried out the trash bags and lamp. Barry put the refrigerator on the hand truck. Then he carefully loaded it onto the pickup.

"What's wrong with that refrigerator?" Chad asked on the way home.

"Looks all right to me," said Barry.

When he got home he put the refrigerator in the garage and plugged it in. It ran fine. Barry looked it over. "It's a little dirty," he said to Chad. "And it needs a new gasket. But I'd say there's 75 dollars in it. I'll tell you what, buddy. You help me fix up this refrigerator and I'll split whatever we get for it. What do you say to that deal?"

"Cool!" said Chad.

"And all this lamp needs is a new cord," Barry added.

"I just decided what I want to be when I grow up," said Chad.

"What's that, buddy?"

"A *hauler*!" said Chad. "I want to be a hauler—just like my dad."

CHAPTER 7

Money in Junk

Barry soon learned that fine china wasn't the only throw-away that was worth money. He put ads in the newspaper. He got 100 dollars for the cleaned-up refrigerator with a new gasket. He gave 50 to Chad. He paid a dollar for a new cord and got ten dollars for the lamp. He began keeping an eye out for anything at all that he might be able to sell. He looked through everything before taking a load over to the dump.

Barry also began hauling for Allen, the builder. Allen's workplaces were often a mess, but the hard work was worth the trouble. Barry usually got some good scrap wood and metal on those jobs.

Barry hauled away all kinds of old furniture, household goods, and large and small appliances. One time he ran across an old lawn mower. He got rid of the gas in it before he put it in the truck. He didn't want to haul anything that might catch on fire.

Barry set up four big sorting bins in his garage. Then he decided what to do with each item he brought home: fix it, sell it, use it, or give it away. Each type of item went into its own bin.

In the "Fix" bin Barry put toasters, furniture, lamps, and stereos. His worktable and some tools were right next to that bin. Even the worktable and most of the tools were things that people had asked him to get rid of!

In the "Sell" bin he put anything that was ready to go. As soon as he had fixed something, it went in there. He set up three buckets for parts: plumbing, electrical, and auto. He put the scrap wood and metal out back.

In the "Use" bin were things Barry wanted for himself. He kept tools and wood there that he could use later. He even kept things that he could use as gifts. Chad got a bunch of toy cars and trucks for his birthday. Some of them had never been out of the boxes they came in.

In the "Give Away" bin went things that weren't worth fixing but were too good to dump.

The garage filled up fast. At first Barry was too busy to keep up with everything. He spent most of his days hauling—even some Saturdays and Sundays now. He didn't have time to fix all the things he wanted to fix. He wasn't home enough to sell things that he planned to sell. And the "Give Away" bin was starting to look like a trash heap.

If he didn't get some of the things moving soon, he would run out of space. But he didn't want to throw anything away that he might want later. To Barry the garage looked like a pile of dust, covering up a pile of gold.

Then the weather changed. Winter was coming. Not as many people were moving. And not as many people were cleaning out their basements. For the first time in months, Barry wasn't working every day.

The cold weather came just in time. Barry put a space heater in the garage and got to work. He quickly became very busy again. Whenever he didn't have a hauling job, he spent his days fixing and selling the stuff he'd brought home.

Barry got very good at cleaning all sorts of things. He took many coats of paint off of old furniture. He gave some pieces a new coat of paint. Other pieces got wood stain. Some pieces got nothing but a nice coat of wax.

He fixed up the large and small appliances. If he needed to, he bought parts for them. Sometimes he took parts from appliances that could not be fixed. Barry found that he was more of a fix-it person than he knew. He learned a lot along the way, too.

Each week he put a For Sale ad in the newspaper. He sold lots of things from those ads. He took the really special pieces to dealers. They kept ten cents on the dollar for whatever price they got. Now and then he had a big garage sale to get rid of little things.

Once a month Barry took a load of metal and car batteries to the scrap metal yard. He sold buckets full of parts to plumbers, electricians, and auto mechanics.

Finally Barry got around to giving some things away. One day he loaded up the truck with books and clothing. He hauled them to charity groups. He did, however, ask for receipts every time. He saved all the receipt slips so he could take something off his taxes at the end of the year.

One time he dropped off a truck full of clothing at a homeless shelter. The people there thought that their dreams had come true.

By January the garage was almost empty. And the phone was still not ringing off the wall with new work. It was time for a new idea.

CHAPTER 8

Branching Out

"Everything seemed to be going great," Barry said to Melissa. "Now I have next to no work. Maybe I should look for a job."

"Now wait a minute," said Melissa. "Remember what happened when you hurt your back? Don't be willing to give in so quickly. You just have a *temporary* problem here. I think you can solve it."

"It's a problem that could last until spring," Barry said. "That's even worse than when I hurt my back."

"How about your resale business?" Melissa asked. "That was going great."

"I'm not hauling much, so I'm not keeping much."

"Well, then, what's another way you could make money with your truck?" Melissa asked. She got out a paper and pencil. The two of them started talking about it. Melissa wrote down their ideas.

"I was turning down yard work," said Barry. "I guess I could do yard work. You know—rake leaves, trim trees, clean out lots. But most of that's done in the fall, and now it's winter."

"I'll write that down for next year," said Melissa.

"I heard that the big department stores hire guys like you to deliver large appliances," said Melissa.

"OK. I'll check on that. That sounds like a good idea. With my hand truck I have no problem with refrigerators and stoves. By now large appliances are a real piece of cake."

Melissa wrote that one down.

"I could tear down small buildings," said Barry. "But I would need help for that. It's really hard work. I know some haulers who do it. They get a lot of stuff out of those old buildings. They make money selling window frames and wooden beams and stuff like that."

"So . . ." Melissa started.

"I really don't think that's for me," said Barry. "And I couldn't do it anyway—not without insurance."

Then it started to snow.

At first Barry and Melissa didn't see the white flakes falling softly past the window. Then Barry caught a flash of something white out of the corner of his eye. His face lit up.

"That's it!" said Barry. "Snow! What comes down must get picked up."

"What on earth are you talking about?" Melissa asked.

"Anyone who has a driveway must

clear the snow off of it," said Barry. "People have to get their cars out to go to work. Not everyone is willing or able to shovel a whole driveway."

"But you don't have a snow plow."

Barry smiled. "I'll go to the bank and ask for a loan," he said. "If I put the truck up as a guarantee, I can get a loan to buy a snow plow. I bet I could make a lot of money on snow this winter!"

That's just what Barry did. He couldn't buy a snow plow that day. He had to wait for the loan to go through. It snowed twice while he waited. But he went ahead and ordered a good plow. At last the bank called. Barry got the loan and picked up the plow.

Barry learned how to hook up the plow to the truck. He also bought a very good snow shovel. He put a "THINK SNOW" bumper sticker on his truck. And he listened to the weather report on TV every night.

About a week later there was a small

storm. It snowed during the night. Barry woke up at 4:00 A.M. and hooked up the plow. At 6:00 the snow turned to rain. By 8:00 that morning, the snow was nearly gone. Barry took the plow off the truck and went back to bed.

Then, a few days later, a bigger storm came. It started slowly, late in the evening. Barry didn't hook up the plow. It was too much trouble if the snow stopped early. But it didn't stop. It snowed all night.

Barry got up very early the next morning. He hooked up the plow. By 6:00 more than six inches of snow had fallen on the ground.

Chad had slept over that night. Barry went into the boy's room to wake him. "Get ready, buddy," Barry said. "The snow's here and school's called off. You all set to come along and shovel the walks and steps?"

Chad jumped out of bed. In a flash he was dressed and ready to go.

Then Barry and his son set out looking for work. A neighbor up the street was shoveling his driveway. "Need some help?" Barry called out.

"How much do you charge?" said the neighbor. Barry told him. Then the man said, "I've been shoveling my own driveway for 30 years. But I'm not as young as I used to be. You can do it for me this time!"

By noon Barry had plowed ten driveways. Chad had shoveled walks and steps at every house. The little guy was very tired.

After a lunch break, Barry dropped Chad off at Suzanne's. He worked alone that afternoon, plowing and shoveling. By the time it got dark he was feeling almost rich. But he was very, very tired.

CHAPTER 9

Turning Point

It was a long, snowy winter. "Lucky for us," Barry told Chad. "We need all the work we can get!"

Whenever a big storm closed the schools, Chad worked with Barry. Some of the storms came on weekends. So Barry usually had a helper then. Of course, he paid his son. Chad made pretty good money for a boy his age. And he loved going along with his dad.

Then winter was over. The snow gave way to green grass and buds on the trees. Bright little flowers popped out of the ground. Barry put the snow plow in the garage and got the truck ready for spring.

He had almost forgotten about his old job at the plant. He had been laid off for a long time now. Somehow he didn't expect the plant to ever open again.

But one day Barry was getting ready for a hauling job. He had one foot out of the door when the phone rang. He went back inside to pick it up.

"This is Joe Logan from the plant," said the voice on the other end. "We're starting up again in two weeks. We're calling back half of the day shift. Your name is on the list. What do you say?"

Barry stopped in his tracks. He wasn't ready to hear this. He didn't know what to say. There was a time when he might have jumped at the chance. Now he wasn't so sure.

"What's going on, Barry?" said Joe Logan. "Find another job?"

"Actually, I have," said Barry. "I'm in business for myself."

"Doing what?" Joe asked.

"Well, I started a small hauling business," Barry answered. "I resell a lot of things that I pick up. And I plowed snow during the winter."

"Sounds like you've been keeping pretty busy," Joe said.

"I sure have," said Barry.

"Are you actually making a living at this?" Joe wondered.

"Believe it or not, Joe, I am. Some months I make more than I did at the plant," said Barry.

"Well, that sounds great," Joe said. "Of course, this would be a steady paycheck again. What do you say?"

"Give me a day to think about it," said Barry as he hung up the phone.

Barry had two hauling jobs that day. But all he could think about was that phone call from Joe Logan.

He had enjoyed his job at the plant. He wasn't afraid of hard work, that was for sure. It would be great to have every weekend off again. And it would be real nice to get a paycheck every Friday. There was a lot to be said for the old days at the plant.

On the other hand, he would miss working on his own. Every hauling job was a new story. It was fun to meet new people all the time. Why, he had met Melissa on the very first job. The money was good, too—most of the time, anyway. And he never knew what small treasures he might come across. There was more money in that, too.

Barry had dinner with Melissa that night. He told her about the phone call from Joe Logan. He told her what he had been thinking about all day.

"I'd sure get to see more of you if you had an eight-hour job," Melissa said. "That would be nice. But you would really miss this business, wouldn't you? You've already put so much time and effort into it."

"I'd miss it a lot," said Barry. "It's tough and it's dirty. You never know what's happening next. But I've never had so much fun in my life!"

"So what are you going to tell Joe Logan?" Melissa asked.

"I might be the biggest fool in the world," said Barry. "But something inside is telling me to stick with the hauling business."

Melissa took his hand. "I'm behind you all the way."

The next day Barry did a job in the morning. He knew that Joe would be calling at lunchtime. And he also knew what he would say to him. About 11:30 Barry headed home.

The moving job was pretty far out of town. Barry was running late and he didn't want to miss Joe's call. He was driving fast. He never saw the temperature needle go way over into the red. He didn't know how hot the engine was getting. He didn't know there was a hole in the radiator. He didn't see the trail of water he was leaving behind.

When the truck began to jerk, Barry looked down at the dashboard. As soon as he saw what was happening, he pulled off the highway. It was the first time his trusty pickup had broken down since he started hauling.

What a thing to happen *now*! Just when things were going smoothly again. Just when he had made up his mind to stay in business for himself.

C H A P T E R   10

Moving Ahead

Barry didn't get to talk to Joe Logan that day. It took all afternoon to deal with the truck. First he had the truck towed in to a garage. He found out that the radiator work would ground him for nearly a week. Barry felt worried and helpless. Things like this wouldn't matter as much if he worked at the plant.

Barry talked to Joe Logan the next day. He was coming off a good night's sleep and he felt strong and ready for anything. He couldn't wait to get the truck back and start working again. He tried to see himself back at the plant— but he just couldn't.

Barry told Joe Logan, "No, thanks." He sure hoped that he woudn't regret it. But he had to go with the feeling in his gut. And his gut was telling him to stick with his own business.

While the truck was in the shop, Barry got ready for warm weather. He spent the week putting up new signs, placing new ads, talking to new people. He bought himself a beeper. That way he could return phone calls right away, no matter where he was.

Barry was back on the road by the end of the week. In a few days he had a schedule worked out. On Monday and Wednesday mornings he delivered large appliances for a big store at the mall. The rest of the time he did moving and hauling jobs. About every other week, he did a job for Allen, the builder. He saved his nights and weekends for fixing and selling things that people had asked him to take away.

The plan worked well all through the warm months. In the fall Barry took jobs clearing and hauling leaves and brush. Sometimes he even trimmed trees. He saved the good wood and cut it up for firewood. Then he delivered it to anyone who bought it.

At the end of the year, Barry looked over his records. He had been in business for a year and a half. He had written down every penny he spent and every penny he took in. Now he worked out how much he had really made that year. Even after child support and a big repair bill on the truck, the "bottom line" looked very good. All of it had come from his own business.

Then Suzanne got sick. She was out of work and had to stay home in bed. She couldn't take care of Chad. So the boy went to live with Barry for a while.

Barry wasn't used to taking care of a child. He couldn't be home when Chad got home from school. The boy had to be home alone for more than an hour every weekday. Whenever he could, Barry took his son on jobs. And by now Chad enjoyed being with Melissa. But it was a tough time for everyone.

A few months later Suzanne got a little better. She went back to work part-time. But she still wasn't strong enough to take Chad back.

Then there was Melissa. Barry had fallen in love with her. They saw each other every chance they got.

One night Chad was staying at a friend's house. Barry and Melissa were having dinner at a nice restaurant. The music was right. The lights were right. The time was right.

"I didn't think I would ever want to get married again," Barry told Melissa. He took her hand in his. "But I do. I want to marry you. Will you marry me?"

"Yes, I will," Melissa answered. Her face glowed.

They were married a month later. It was a very small wedding. But even Suzanne was there.

While Barry and Melissa took a little trip, Chad stayed with Suzanne.

When they got back, the answering machine was blinking. Barry spent the whole morning returning calls. He lined up plenty of work.

One of the calls was from Joe Logan. "Guess what?" he said. "You made the right move. Now *I'm* out of a job. The company decided to take the whole plant out of the country. Can you believe it?"

The next time Barry needed someone to help him, he called Joe Logan. He was glad to help out an old friend. Life was good for Barry. Now and then he had a slow day. But those days were few. There always seemed to be plenty of work for a man and his pickup.